Ajesh Faizal

Microprocessors and Microcontrollers Laboratory Hand Book

Ajesh Faizal

Microprocessors and Microcontrollers Laboratory Hand Book

Noor Publishing

Imprint

Any brand names and product names mentioned in this book are subject to trademark, brand or patent protection and are trademarks or registered trademarks of their respective holders. The use of brand names, product names, common names, trade names, product descriptions etc. even without a particular marking in this work is in no way to be construed to mean that such names may be regarded as unrestricted in respect of trademark and brand protection legislation and could thus be used by anyone.

Cover image: www.ingimage.com

Publisher:
Noor Publishing
is a trademark of
International Book Market Service Ltd., member of OmniScriptum Publishing Group
17 Meldrum Street, Beau Bassin 71504, Mauritius
Printed at: see last page
ISBN: 978-620-2-79002-4

MICROPROCESSORS AND MICROCONTROLLERS LABORATORY HAND BOOK

AJESH FAIZAL

Professor, Department of Computer Science and Engineering, Musaliar College of Engineering and Technology, Pathanamthitta, India

ACKNOWLEDGEMENT

***The endless thanks go to The Lord Almighty for all the blessings he has showered onto us,
process of putting this book***

*We sincerely offer deepest gratitude and thanks **Mr P I Sherief Muhammedh**, Chairman,
Group of Institutions, **Prof.JayaPrasad R**, Dean, Musaliar College of Engineering, and
their constant guidance and encouragement to accomplish this
book.*

*I am extremely thankful to my wife Mrs. **Shibina M R** and my son **Aamil Mehboob** for their
care and continuous encouragement throughout the preparation of this book. I solemnly thank my father
Mr. M Faizal and mother Mrs. N Seenath , for their encouragement to start this work*

*Also, we would like to thank our colleagues for their support and encouragement that they have
given us. It provide us with a great opportunity to look back, and thank to all those who have
been directly or indirectly instrumental in successful completion of this book.*

CHAPTER I
INTRODUCTION
1.1 8086 MICROPROCESSOR

PIN DIAGRAM

	8086		MAX MODE	MIN MODE
Vss (GND)	1	40	Vcc (5P)	
AD14	2	39	AD15	
AD13	3	38	A16/S3	
AD12	4	37	A17/S4	
AD11	5	36	A18/S5	
AD10	6	35	A19/S6	
AD9	7	34	$\overline{BHE}$/S7	
AD8	8	33	MN/$\overline{MX}$	
AD7	9	32	$\overline{RD}$	
AD6	10	31	$\overline{RQ/GT0}$	HOLD
AD5	11	30	$\overline{RQ/GT1}$	HLDA
AD4	12	29	$\overline{LOCK}$	$\overline{WR}$
AD3	13	28	$\overline{S2}$	M/$\overline{IO}$
AD2	14	27	$\overline{S1}$	DT/$\overline{R}$
AD1	15	26	$\overline{S0}$	$\overline{DEN}$
AD0	16	25	QS0	ALE
NMI	17	24	QS1	$\overline{INTA}$
INTR	18	23	$\overline{TEST}$	
CLK	19	22	READY	
Vss (GND)	20	21	RESET	

1.2 ARCHITECTURAL DIAGRAM OF 8086

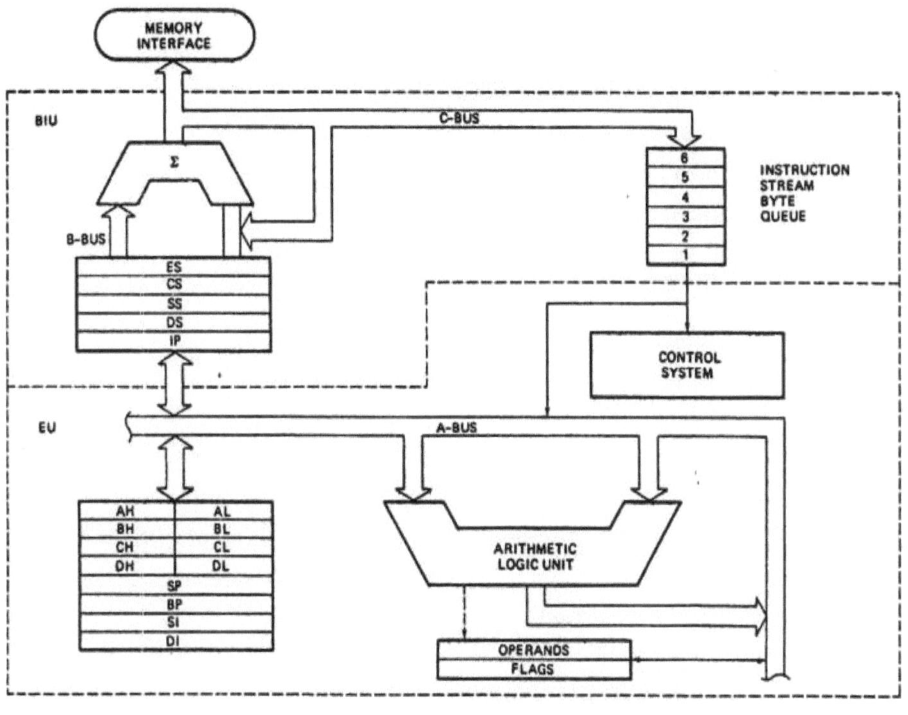

1.3 Review Questions:

1. What are the flags in 8086?
2. What are the various interrupts in 8086?
3. What is meant by Maskable interrupts?
4. What is Non-Maskable interrupts?
5. Which interrupts are generally used for critical events?
6. Give examples for Maskable interrupts?
7. Give example for Non-Maskable interrupts?
8. What is the Maximum clock frequency in 8086?
9. What are the various segment register in 8086?
10. Which Stack is used in 8086?
11. What are the address lines for the software interrupts?
12. What is a SIM and RIM instruction?
13. Which is the tool used to connect the user and the computer?
14. What is the position of the Stack Pointer after the PUSH instruction?
15. What is the position of the Stack Pointer after the POP instruction?
16. Logic calculations are done in which type of registers?
17. What are the different functional units in 8086?
18. Give examples for Microcontroller?
19. What is meant by cross-compiler?
20. What are the address lines for the hardware interrupts?
21. Which Segment is used to store interrupt and subroutine return address registers?
22. Which Flags can be set or reset by the programmer and also used to control the operation of the processor?
23. What does EU do?
24. Which microprocessor accepts the program written for 8086 without any changes?
25. What is the difference between 8086 and 8088?

<u>1.4 8051 MICROCONTROLLER</u>

INTRODUCTION TO MICROCONTROLLER 8051

- A smaller computer
- On-chip RAM, ROM, I/O ports...
- Example: Motorola's 6811, Intel's 8051, Zilog's Z8 and PIC 16X

<u>BLOCK DIAGRAM OF 8051</u>

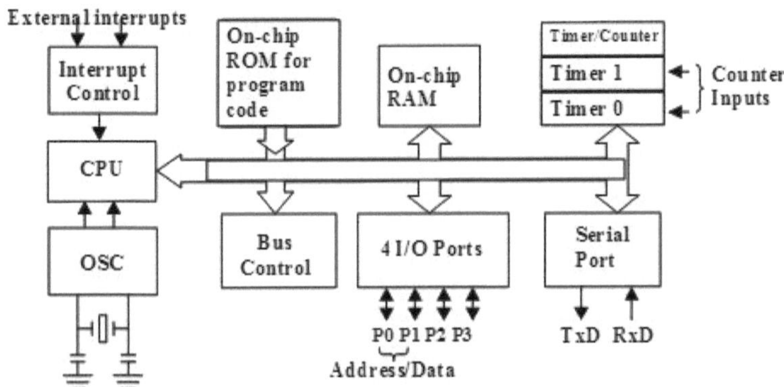

<u>PIN DIAGRAM OF 8051</u>

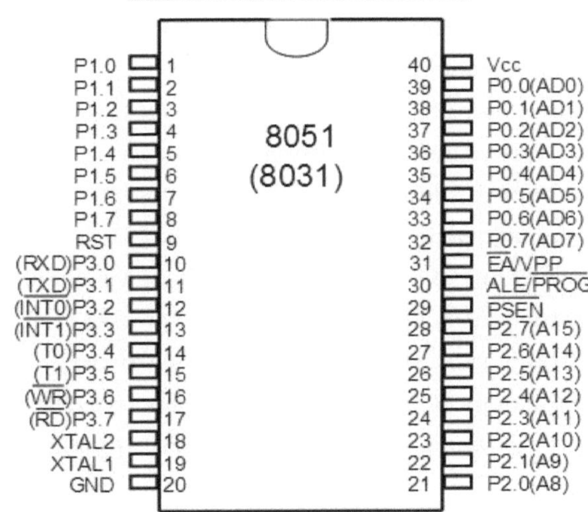

1.5 COMPARISON BETWEEN MICROPROCESSOR AND MICROCONTROLLER

Microprocessor	Microcontroller
<ul><li>CPU is stand-alone, RAM, ROM, I/O, timer are separate</li><li>Designer can decide on the amount of ROM, RAM and I/O ports.</li><li>Expansive</li><li>Versatility</li><li>General-purpose</li></ul>	<ul><li>CPU, RAM, ROM, I/O and timer are all on a single chip.</li><li>Fix amount of on-chip ROM, RAM and I/O ports.</li><li>For applications in which cost, power and space are critical</li><li>Single-purpose</li></ul>

Review Questions:
1. What is a state in 8031/8051 microcontroller?
2. How many machine cycles are needed to execute an instruction in 8031/8051 controller?
3. How to estimate the time taken to execute an instruction in 8031/8051 controller?
4. What is the size of 8031/8051 instructions?
5. List the various machine cycles of 8031/8051 controller.
6. How the 8051 microcontroller differentiates external program memory access and data memory access?
7. What are the addressing modes available in 8051 Controller?
8. Explain the register indirect addressing in 8051.
9. Explain the relative addressing in 8051.
10. How the 8051 instructions can be classified?
11. List the instructions of 8051 that affect all the flags of 8051.
12. List the instructions of 8051 that affect overflow flag in 8051.
13. List the instructions of 8051 that affect only carry flag.
14. List the instructions of 8051 that always clear carry flag.
15. What are the operations performed by Boolean variable instructions of 8051?

CHAPTER II

PROGRAMMING WITH 8086

2.1   16-BIT DATA ADDITION AND SUBTRACTION

PROGRAM:

ADDITION

```
MOV SI, 1100ᴴ
MOV AX, [SI]
MOV BX, [SI+2]
MOV CL, 00ᴴ
ADD AX, BX
MOV [SI+4], AX
JNC 0113
INC CL
MOV [SI+6], CL
INT A5
```

PROGRAM:

SUBTRACTION

```
MOV SI, 1100ᴴ
MOV AX, [SI]
MOV BX, [SI+2]
MOV CL, 00ᴴ
SUB AX, BX
MOV [SI+4], AX
JNC 0113
INC CL
MOV [SI+6], CL
INT A5
```

<u>2.1   16-BIT DATA ADDITION AND SUBTRACTION</u>

AIM:

To add and subtract two 16-bit numbers stored at consecutive memory locations.

APPARATUS REQUIRED: 8086 Microprocessor Kit.

ALGORITHM:
1. Load the first data in AX register.
2. Load the second data in BX register.
3. Clear CL register.
4. Add/subtract the two data and get the sum in AX register.
5. Store the sum in memory.
6. Check for carry, if carry flag is set then go to next step, otherwise go to step-8.
7. Increment CL register.
8. Store the carry in memory.
9. Stop.

OBSERVATION:

ADDITION: WITH CARRY

Input Data:	Output Data:
1100 – 7C	1104 – 21
1101 – 87	1105 – 42
1102 – A5	1106 – 01
1103 – BA	

ADDITION: WITHOUT CARRY:

Input Data:	Output Data:
1100 – CA	1104 – ED
1101 – 05	1105 – A6
1102 – 23	1106 – 00
1103 – A1	

SUBTRACTION: WITH BORROW

Input Data:	Output Data:
1100 – A1	1104 – FE
1101 – 63	1105 – 2D
1102 – A3	1106 – 00
1103 – 35	

SUBTRACTION: WITHOUT BORROW

Input Data:	Output Data:
1100 – BC	1104 – F1
1101 – 57	1105 – E1
1102 – CB	1106 – 01
1103 – 65	

RESULT:

Thus the two 16-bit numbers are added, subtracted and results are stored at memory locations.

PROGRAM:

```
MOV SI, 1100
MOV DI, 1201
MOV BP, 1301
MOV CL, [SI]
INC SI
MOV BX, 00H
MOV DL, 00H
CLC
MOV AL, [SI+BX]
ADC/SBB AL, [DI+BX]
MOV [BP], AL
INC BX
INC BP
LOOP 0112
JNC 0121
INC DL
MOV [BP], DL
INT A5
```

ADDITION:

Input Data:		Output Data:
1100 – 06		
1101 – 07	1201 – 04	1301 – 0B
1102 – 03	1202 – 07	1302 – 0A
1103 – 01	1203 – 02	1303 – 03
1104 – 03	1204 – 04	1304 – 07
1105 – 02	1205 – 04	1305 – 06
1106 – 04	1206 – 02	1306 – 06
		1307 – 00

2.2. MULTI-BYTE ADDITION AND SUBTRACTION

AIM:

To add and subtract two multi-byte data.

APPARATUS REQUIRED: 8086 Microprocessor Kit.

ALGORITHM:

1. Load the starting address of 1st data in SI register.
2. Load the starting address of 2nd data in DI register.
3. Load the starting address of result in BP register.
4. Load the byte count in CL register.
5. Let BX register be byte pointer, initialize byte pointer as zero.
6. Clear DL register to account for final carry.
7. Clear carry flag.
8. Load a byte of 1st data in AL register.
9. Add /subtract the corresponding byte of 2nd data in memory to AL register along with previous carry.
10. Store the sum in memory.
11. Increment the byte pointer (BX) and result pointer (BP).
12. Decrement the byte count (CL).
13. If byte count (CL) is zero go to next step, otherwise go to step-8.
14. Check for carry. If carry flag is set then go to next step, otherwise store the data.
15. Increment the DL register.
16. Store final carry in memory.
17. Stop

OBSERVATION:

SUBTRACTION:

Input Data:		Output Data:
1100 – 06		
1101 – 08	1201 – 02	1301 - 06
1102 – 07	1202 – 03	1302 - 04
1103 – 0A	1203 – 02	1303 - 08
1104 – 0E	1204 – 0A	1304 - 04
1105 – 02	1205 – 01	1305 – 01
1106 – 0F	1206 – 02	1306 - 0D
		1307 - 00

RESULT:

Thus the Multi-byte addition and Multi-byte subtraction are performed.

<u>**2.3. BCD ADDITION AND SUBTRACTION**</u>

PROGRAM:

```
MOV SI, 1100
MOV CL, 00
MOV AX, [SI]
MOV BX, [SI+2]
ADD/SUB AL, BL
DAA/DAS
MOV DL, AL
MOV AL, AH
ADC/SBB AL, BH
DAA/DAS
MOV DH, AL
JNC 011A
INC CL
MOV [SI+4], DX
MOV [SI+6], CL
INT A5
```

SUBTRACTION:

Input Data:	Output Data:
1100 – 07	1104 – 03
1101 – 08	1105 – 05
1102 – 04	1106 – 00
1103 – 03	

2.3. BCD ADDITION AND SUBTRACTION

AIM:

To add and subtract two numbers of BCD data.

APPARATUS REQUIRED: 8086 Microprocessor Kit.

ALGORITHM:

1. Load the address of data in SI register.
2. Clear CL register to account for carry.
3. Load the first data in AX register and second data in BX register.
4. Perform binary addition/subtraction of low byte of data to get the sum/difference in AL register.
5. Adjust the sum/ difference of low bytes to BCD.
6. Save the sum/ difference of low bytes in DL register.
7. Get the high byte of first data in AL register.
8. Add/sub the high byte of second data and previous carry to AL register. Now the sum of high bytes will be in AL register.
9. Adjust the sum/ difference of high bytes to BCD.
10. Save the sum/ difference of high bytes in DH register.
11. Check for carry. If carry flag is set then go to next step, otherwise go to next condition.
12. Increment CL register.
13. Save the sum /difference (DX register) in memory.
14. Save the carry (CL register) in memory.
15. Stop.

OBSERVATION:

ADDITION:

Input Data:	Output Data:
1100 – 67	1104 – 22
1101 – 74	1105 – 18
1102 – 55	1106 – 01
1103 – 43	

RESULT:

Thus the addition and subtraction of two BCD numbers was performed.

2.4. 16-BIT MULTIPLICATION AND DIVISION

PROGRAM:

```
MOV SI, 1100H
MOV AX, [SI]
MOV BX, [SI+2]
MUL/DIV BX
MOV [SI+4], AX
MOV [SI+6], DX
INT A5
```

2.4. 16-BIT MULTIPLICATION AND DIVISION

AIM:

To multiply/divide two 16-bit data.

APPARATUS REQUIRED: 8086 Microprocessor Kit.

ALGORITHM:
1. Load the address of data in SI register.
2. Get the first data in AX register.
3. Get the second data in BX register.
4. Multiply/divide the content of AX and BX. The product will be in AX and DX.
5. Save the result (AX and DX) in memory. For division the quotient is stored in AX and the remainder in DX.
6. Stop.

OBSERVATION:

MULTIPLICATION:

Input Data:	Output Data:	
1100 – 3A	1104 – 6E	PRODUCT
1101 – 58	1105 – ED	
1102 – 63	1106 – 5E	OVERFLOW
1103 – 15	1107 - 07	

DIVISION:

Input Data:		Output Data:	
1100 – CB	DIVIDEND	1104 – 02	QUOTIENT
1101 – C2		1105 – 00	
1102 – 21	DIVISOR	1106 – 81	REMINDER
1103 – 63		1107 - 07	

Input Data:		Output Data:	
1100 – 24	DIVIDEND	1104 – 00	QUOTIENT
1101 – 48		1105 – 00	
1102 – 87	DIVISOR	1106 – 24	REMINDER
1103 – 78		1107 – 48	

RESULT:

Thus the multiplication and division of two 16-bit numbers was performed.

2.5. LOGICAL OPERATIONS

PROGRAM:

OR/XOR/AND
```
MOV SI, 2000
MOV AX, [SI]
MOV BX, [SI+2]
OR/XOR/AND AX, BX
MOV [SI+4], AX
INT A5
```

NOT
```
MOV SI, 2000H
MOV AX, [SI]
NOT AX
MOV [SI+4], AX
INT A5
```

OBSERVATION:

OR:

Input Data:	Output Data:
2000 – 01	2004 – 11
2001 – 11	2005 – 11
2002 – 10	
2003 – 00	

XOR:

Input Data:	Output Data:
2000 – 00	2004 – 01
2001 – 11	2005 – 01
2002 – 01	
2003 – 10	

AND:

Input Data:	Output Data:
2000 – 00	2004 – 10
2001 – 11	2005 – 00
2002 – 01	
2003 – 10	

NAND:

Input Data:	Output Data:
2000 – 01	2004 – FF
2001 – 10	2005 - EF
2002 – 00	
2003 – 11	

<h1 style="text-align:center"><u>2.5. LOGICAL OPERATIONS</u></h1>

AIM:

To perform logical operations (OR/XOR/AND) on 16-bit data using 8086 microprocessor.

APPARATUS REQUIRED: 8086 Microprocessor Kit.

ALGORITHM:

1. Set the Source Index address as 2000.
2. Load the data in address of SI to AX.
3. Load the Second data in address of [SI+02] to BX.
4. Perform OR/XOR/AND operation on AX and BX.
5. Perform NOT Operation on AX Register.
6. Move the Content of AX to [SI+04] address.
7. Stop.

OBSERVATION:

NOR:

Input Data:	Output Data:
2000 - 01	2004 - EE
2001 - 10	2005 - EE
2002 - 11	
2003 - 01	

XNOR:

Input Data:	Output Data:
2000 - 11	2004 - EF
2001 - 00	2005 - FF
2002 - 01	
2003 - 10	

NOT:

Input Data:	Output Data:
2000 - 11	2004 - EE
2001 - 10	2005 - EF

RESULT:

Thus the logical operation on 16 bit data was performed.

2.6. MOVING A BLOCK OF DATA

PROGRAM:

LABEL	MNEMONICS	COMMENTS
	ASSUME CS:CODE,DS:CODE	SET CODE SEGMENT AND DATA SEGMENT AS CODE
	CODE SEGMENT	START CODE SEGMENT
	ORG 1000$_H$	SET THE STARTING ADDRESS OF THE PROGRAM
	MOV SI, 1400$_H$	SET SOURCE INDEX ADDRESS AS 1400
	MOV CL, [SI]	MOVE THE COUNT VALUE FROM SI ADDRESS TO CL
	MOV SI, 1500$_H$	SET SOURCE INDEX ADDRESS AS 1500
	MOV DI, 1600$_H$	SET DESTINATION INDEX ADDRESS AS 1600
L1:	MOV AX, [SI]	MOVE THE DATA FROM SI ADDRESS TO AX
	MOV [DI], AX	MOVE THE DATA FROM AX TO DI ADDRESS
	ADD SI, 02$_H$	ADD 02H WITH ADDRESS IN SI
	ADD DI, 02$_H$	ADD 02H WITH ADDRESS IN DI
	DEC CL	DECREMENT CL
	JNZ L1	CHECK CL=0 IF NO JUMP TO L1
	MOV AH,4CH	EXIT TO DOS
	INT 21H	
	CODE ENDS	CODESEGMENT END
	END	END OF THE PROGRAM

2.6. MOVING A BLOCK OF DATA

AIM:

To move a block of data from Address Location 1500 to Address Location 1600 using 8086 MASM Software.

APPARATUS REQUIRED: 8086 Microprocessor Kit.

ALGORITHM:

1. Load the address of data in SI register.
2. Get the count value from SI address to CL.
3. Load the address of data in SI register.
4. Load the address of data in DI register.
5. Move data from SI address to AX.
6. Move data from AX to SI address.
7. Add 02 with address in SI and DI.
8. Decrement Count Value.
9. Stop the program.

OBSERVATION:

ADDITION:

Input Data:		Output Data:
1400 – 05	(COUNT)	
1500 – 06		1600 - 01
1501 – 07		1601 - 09
1502 – 08		1602 - 08
1503 – 09		1603 - 07
1504 - 01		1604 - 06

RESULT:

Thus the moving block of data from 1500 to 1600 using 8086 MASM software was performed.

2.7. BCD TO HEXADECIMAL CONVERSION

PROGRAM:

```
MOV BX, 1100H
MOV AL, [BX]
MOV DL, AL
AND DL, 07H
AND AL, 70H
MOV CL, 04H
ROR AL, CL
MOV DH, 0AH
MUL DH
ADD AL, DL
MOV [BX+1], AL
INT A5
```

BCD TO HEXADECIMAL CONVERSION

AIM:

To convert a given BCD data into Hexadecimal data.

APPARATUS REQUIRED: 8086 Microprocessor Kit.

ALGORITHM:

1. Load the address of BCD data in BX register.
2. Get the BCD data in AL register.
3. Copy the BCD data in DL register.
4. Logically AND DL with 07H to mask upper nibble and get units digit in DL.
5. Logically AND AL with 70H to mask lower nibble.
6. Move the count value for rotation in CL register.
7. Rotate the content of AL to move the upper nibble to lower nibble position.
8. Move 0AH to DH register.
9. Multiply AL with DH register. The product will be in AL register.
10. Add the unit digit in DL register to product in AL register.
11. Save the binary data (AL) in memory.
12. Stop.

OBSERVATION:

Input Data:	**Output Data:**
1101 – 18	1101 – 0A

RESULT:

Thus the BCD to Hexadecimal conversion was performed.

2.8. HEXADECIMAL TO BCD CONVERSION

PROGRAM:

```
        MOV BX, 1100H
        MOV AL, [BX]
        MOV DX, 0000H
HUND:   CMP AL, 64H
        JC TEN
        SUB AL, 64H
        INC DL
        JMP HUND
 TEN:   CMP AL, 0AH
        JC UNIT
        SUB AL, 0AH
        INC DH
        JMP TEN
UNIT:   MOV CL, 04
        ROL DH, CL
        ADD AL, DH
        MOV [BX+1], AL
        MOV [BX+2], DL
        INT A5
```

2.8. HEXADECIMAL TO BCD CONVERSION

AIM:

To convert an 8-bit hexadecimal data into BCD.

APPARATUS REQUIRED: 8086 Microprocessor Kit.

ALGORITHM:

1. Load the address of data in BX register.
2. Get the binary data in AL register.
3. Clear DX register for storing hundreds and tens.
4. Compare AL with 64_H (100_{10}).
5. Checks carry flag. If carry flag is set then go to step-9, otherwise go to next step.
6. Subtract 64_H (100_{10}) from AL register.
7. Increments hundreds register (DL).
8. Go to step-4.
9. Compare AL with $0A_H$ (10_{10}).
10. Checks carry flag. If carry flag is set then go to step-14, otherwise go to next step.
11. Subtract $0A_H$ (10_{10}) from AL register.
12. Increments tens register (DH).
13. Go to step-9.
14. Move the count value 04H for rotation in CL register.
15. Rotate the content of DH four times.
16. Add DH to AL to combine tens and units as 2-digit BCD.
17. Save AL and DL in memory.
18. Stop.

OBSERVATION:

Input Data:	**Output Data:**
1100 – 39	1101 - 09

RESULT:

Thus the hexadecimal to BCD conversion was performed.

2.9. HEXADECIMAL TO ASCII CONVERSION

PROGRAM:

```
        MOV SI, 2000H
        MOV DI, 3000H
        MOV AL, [SI]
        SUB AL, 30H
        CMP AL, 09H
        JE L1
        JC L1
        SUB AL, 07H
L1:     MOV [DI], AL
```

2.9. HEXADECIMAL TO ASCII CONVERSION

AIM:

To convert the given Hexadecimal data into ASCII.

APPARATUS REQUIRED: 8086 Microprocessor Kit.

ALGORITHM:

1. Set SI as pointer for binary array.
2. Set DI as pointer for ASCII array.
3. Load the binary into AL
4. Subtract 30H from AL register
5. Compare 09H to AL register if equal go L1
6. In that load the AL register into DI
7. checks carry, if carry flag is set got to L1
8. Subtract 07H to AL register

OBSERVATION:

Input Data:	**Output Data:**
1100 – 41	1400 – 0A

RESULT:

Thus the hexadecimal to ASCII conversion was performed.

2.10 ASCII TO HEXADECIMAL CONVERSION

PROGRAM:

```
        MOV SI, 2000H
        MOV DI, 3000H
        MOV AL, [SI]
        CMP AL, 09H
        JE L1
        JC L1
        ADD AL, 07H
L1:     ADD AL, 30H
        MOV [DI], AL
```

2.10 ASCII TO HEXADECIMAL CONVERSION

AIM:

To convert an ASCII number into hexadecimal.

APPARATUS REQUIRED: 8086 Microprocessor Kit.

ALGORITHM:

1. Set SI as pointer for ASCII array.
2. Set DI as pointer for binary array.
3. Get a byte of ASCII array in AL registers.
4. Compare AL with 09H.
5. If it is equal go to step L1, otherwise go to step 4.
6. Check carry flag, if carry flag is set, go to step L1
7. Add 07H from AL register.
8. Add 07H from AL register.
9. Load the binary array in AL to DI

OBSERVATION:

Input Data:	**Output Data:**
1100 – 0A	1400 – 41

RESULT:

Thus the ASCII to hexadecimal conversion was performed.

2.11.ADDITION/SUBTRACTION OF TWO nxn MATRICES

PROGRAM:

LABEL	MNEMONICS	COMMENTS
	ASSUME CS:CODE,DS:CODE	SET CODE SEGMENT AND DATA SEGMENT AS CODE
	CODE SEGMENT	START CODE SEGMENT
	ORG 1000$_H$	SET THE STARTING ADDRESS OF THE PROGRAM
	MOV SI,2000$_H$	SET SOURCE INDEX ADDRESS AS 2000
	MOV DI,3000$_H$	SET DESTINATION INDEX ADDRESS AS 3000
	MOV CL,04$_H$	SET THE COUNT VALUE 04H IN CL
L1:	MOV AL,[SI]	MOVE THE DATA FROM SI ADDRESS TO AL
	MOV BL,[DI]	MOVE THE DATA FROM DI ADDRESS TO BL
	ADD/SUB AL,BL	ADD/SUBTRACT THE DATA BL FROM AL
	MOV [DI],AL	MOVE THE DATA FROM AL TO DI ADDRESS
	INC DI	INCREMENT DI
	INC SI	INCREMENT SI
	DEC CL	DECREMENT CL
	JNZ L1	CHECK CL=0 IF NO JUMP TO L1
	MOV AH,4C$_H$	EXIT TO DOS
	INT 21$_H$	
	CODE ENDS	CODE SEGMENT END
	END	END OF THE PROGRAM

<h1 style="text-align:center">2.11.ADDITION/SUBTRACTION OF TWO nxn MATRICES</h1>

AIM:

To subtract two nxn matrices using 8086 MASM Software.

APPARATUS REQUIRED: 8086 Microprocessor Kit.

ALGORITHM:

1. Set the Count Value as 04.
2. Set SI register as Pointer for array for 1st Matrix.
3. Set DI register as Pointer for array for 2nd Matrix.
4. Move Contents of SI register to AL.
5. Move contents of DI register to BL.
6. Add/Subtract the contents of AL and BL.
7. Move the result from AL register to Memory.
8. Increment SI and DI register.
9. Decrement Count Value.
10. If Count Value becomes zero, stop the program.

OBSERVATION:

ADDITION:

Input data:		Output data:
2000 – 01	3000 – 04	3000 – 05
2001 – 05	3001 – 02	3001 – 07
2002 – 03	3002 – 07	3002 – 0A
2003 – 04	3003 - 02	3003 – 06

SUBTRACTION:

Input data:		Output data:
2000 – 08	3000 – 04	3000 – 04
2001 – 07	3001 – 03	3001 – 04
2002 – 04	3002 – 02	3002 – 02
2003 – 03	3003 – 01	3003 – 02

RESULT:

Thus the addition and subtraction of two nxn matrices using 8086 MASM software was performed.

<u>2.12 COPY A STRING</u>

PROGRAM:

```
        MOV SI, 2000H
        MOV DI, 2100H
        MOV CX, 04H
        CLD
NEXT:   MOVSB
        LOOP NEXT
        INT A5
```

<u>**2.12 COPY A STRING**</u>

AIM:

To copy a string of data using 8086.

APPARATUS REQUIRED: 8086 Microprocessor Kit.

ALGORITHM:

1. Set SI as pointer for input array of data's.
2. Set DI as pointer for output array of data's.
3. Set CX as counter.
4. Clear direction flag (DF) for auto increment of pointers.
5. Move the string of data's up to the counter value.

OBSERVATION:

Input Data:	**Output Data:**
2000 – 07	2100 – 07
2001 – 03	2101 – 03
2002 – 02	2102 – 02
2003 – 01	2103 – 01

RESULT:

Thus the copy a string of data was performed.

2.13. REVERSE A STRING

PROGRAM:

```
        MOV SI, 2000H
        MOV DI, 3000H
        MOV CX, 0008H
        ADD SI, 07
L1:     MOV AL, [SI]
        MOV [DI], AL
        DEC SI
        INC DI
        DEC CX
        JNZ L1
        INT A5
```

<h1 align="center"><u>2.13.REVERSE A STRING</u></h1>

AIM:

To reverse the input string using 8086 Microprocessor.

APPARATUS REQUIRED: 8086 Microprocessor Kit.

ALGORITHM:

1. Set SI as pointer for input array of data's.
2. Set DI as pointer for output array of data's.
3. Set CX as Counter
4. Add counter of SI with 07
5. Move contents of SI address to AL and from AL TO DL address.
6. Decrement SI and DI to reverse.
7. Decrement CX and report loop if DL to 0
8. Stop the Program.

OBSERVATION:

Input data:	**Output data:**
2000 – 08	3000 – 01
2001 – 07	3001 – 02
2002 – 06	3002 – 03
2003 – 05	3003 – 04
2004 – 04	3004 – 05
2005 – 03	3005 – 06
2006 – 02	3006 – 07
2007 – 01	3007 - 08

RESULT:

Thus the string was reversed using 8086 Microprocessor.

2.14. ASCENDING AND DESCENDING ORDER

PROGRAM:

```
     START: MOV SI, 1100H
            MOV CL, [SI]
            DEC CL
    REPEAT: MOV SI, 1100H
            MOV CH, [SI]
            DEC CH
            INC SI
    REPCOM:  MOV AL, [SI]
            INC SI
            CMP AL, [SI]
            JC/JNC AHEAD
            XCHG AL, [SI]
            XCHG AL, [SI - 1]
     AHEAD: DEC CH
            JNZ REPCOM
            DEC CL
            JNZ REPEAT
            INT A5
```

2.15. ARRANGE AN ARRAY OF DATA IN ASCENDING AND DESCENDING ORDER

AIM:

To sort the given number in the ascending and descending order using 8086 microprocessor.

APPARATUS REQUIRED: 8086 Microprocessor Kit.

ALGORITHM:

1. Set SI register as pointer for array.
2. Set CL register as count for N – 1 repetition.
3. Initialize array pointer.
4. Set CH as count for N – 1 Comparisons.
5. Increment the array pointer.
6. Compare the next element of the array with AL.
7. Check the carry flag. If carry flag is set then go to step-12, otherwise go to next step.
8. Exchange the content of memory pointed by SI and the content of previous memory location
   (For this, exchange AL and memory pointed by SI, and then exchange AL and memory pointed by SI – 1).
9. Decrement the count for comparisons (CH register).
10. Check zero flag. If zero flag is reset then go to step-6, otherwise go to next step.
11. Decrement the count for repetitions (CL register).
12. Check zero flag. If zero flag is reset then go to step-3, otherwise go to next step.
13. Stop.

OBSERVATION:

ASCENDING ORDER

Input data:	Output data:
1100 – 04 (COUNT)	
1101 – 19	1101 – 19
1102 – 43	1102 – 30
1103 – 30	1103 – 43
1104 – 65	1104 – 65

DESCENDING ORDER

Input data:	Output data:
1100 – 04 (COUNT)	
1101 – 03	1101 – 65
1102 – 30	1102 – 40
1103 – 65	1103 – 30
1104 – 40	1104 – 03

RESULT:

Thus the ascending and descending order program is executed and thus the numbers are arranged in ascending order.

PROGRAM:

```
START:      MOV SI, 1100H
            MOV DI, 1200H
            MOV DL, [DI]
            MOV BL, 01H
            MOV AL, [SI]
AGAIN:      CMP AL, DL
            JZ AVAIL
            INC SI
            INC BL
            MOV AL, [SI]
            CMP AL, 20H
            JNZ AGAIN
            MOV CX, 0000H
            MOV [DI+1], CX
            MOV [DI+3], CX
            JMP OVER
AVAIL:      MOV BH, 0FFH
            MOV [DI+1], BH
            MOV [DI+2], BL
            MOV [DI+3], SI
OVER:       INT A5
```

AIM:

To search a given data in an array. Also determine the position and address of the data in the array.

APPARATUS REQUIRED: 8086 Microprocessor Kit.

ALGORITHM:

1. Set SI register as pointer for the array.
2. Set DI register as pointer for given data and result.
3. Get the data to search in DL register.
4. Let BL register keep track of position. Initialize the position count as one.
5. Get an element of array in AL.
6. Compare an element of array (AL) with given data (DL).
7. Check for Zero flag. If zero flag is set then go to step-14, otherwise go to next step.
8. Increment the array pointer (SI) and position count (BL).
9. Get next element of array in AL register.
10. Compare AL with end marker (20_H).
11. Check zero flag. If zero flag is not set, then go to step-6,
12. Otherwise go to next step.
13. Clear CX register and store CX register in four
14. Consecutive locations in memory after the given data.
15. Jump to end (Step-17).
16. Move FF_H to BH register and store it in memory.
17. Store the position count (BL) in memory.
18. Store the address (SI) in memory.
19. Stop.

OBSERVATION:

Input Data:

1100 – 01   1200 – A0 **[Data to Search]**
1101 – 22
1102 – A0
1103 – 45
1104 – 3A
1105 – 20 [End of Array]

Output Data:

1201 – FF **[Data Availability]**
1202 – 03 **[Position]**
1203 – 02 **[LSB of Address]**
1204 – 11 **[MSB of Address]**

RESULT:

Thus searching a given data in an array was performed.

2.17. PASSWORD CHECKING PROGRAM

PROGRAM:

```
        ASSUME CS:CODE, DS:CODE
        DATA SEGMENT
        STRING DB "PANIMALAR"
        BUFFER DB 0FH DUP (?)
        MESSAGE DB 0AH, 0DH, "SORRY!", 0AH, 0DH, "$"
        DATA ENDS
        CODE SEGMENT
        ORG 1000H
START:      MOV AX, DATA
            MOV DS, AX
WAIT:       MOV CX, 09H
            MOV DI, OFFSET BUFFER
NXTCHAR: MOV AH, 08H
            INT 21H
            CMP AL, 0DH
            JE STOP
            MOV [DI], AL
            INC DI
            DEC CX
            JNZ NXTCHAR
STOP:       MOV CX, 09H
            MOV SI, OFFSET STRING
            MOV DI, OFFSET BUFFER
            MOV AX, SEG BUFFER
            MOV ES, AX
            REP   CMPSB
            JNZ SORRY
            MOV AH, 4CH
            INT 21H
SORRY:      MOV DX, OFFSET MESSAGE
            MOV AH, 09H
            INT 21H
            JMP WAIT
            CODE ENDS
            END START
```

2.17. PASSWORD CHECKING PROGRAM

AIM:

To read password and validate the user.

APPARATUS REQUIRED: 8086 Microprocessor Kit.

ALGORITHM:

1. Start the Program.
2. Initialize data Segment.
3. Display Message.
4. Clear Count and number of match.
5. Check if count=9 if not continue.
6. Road and save Character.
7. Restore Character set pointer to password.
8. Compare read character with password.
9. Increment match count if match occurs.
10. Increment pointer and counter.
11. If match count=9 display Message.
12. If match count<>9 display Message SORRY.
13. Stop the program.

RESULT:

Thus the password checking program was implemented and verified successfully.

<u>**2.18. TIME DELAY**</u>

PROGRAM:

```
        MOV DX, 0000
        INT AC
LOOP1:  INT AB
        MOV AL, 04
        INT AE
        MOV AL, DL
        ADD AL, 01
        DAA
        MOV DL, AL
        MOV AL, DH
        ADC AL, 00
        DAA
        MOV DH, AL
        MOV CX, 4000 (DELAY COUNT)
        INT AA
        MOV AH, 0B
        INT A1
        CMP AL, 00
        JZ LOOP1
        MOV AH, 08
        INT A1
        CMP AL, 46  (F→ASCII)
        JZ LOOP2
        CMP AL, 1B
        JNZ LOOP1
        INT A3
LOOP2:  MOV AH, 08
        INT A1
        CMP AL, 53   (S→ASCII)
        JZ LOOP1
        JMP LOOP2
        INT A5
```

<u>NOTE:</u>
F → TO PAUSE
S→ TO CONTINUE
ESC→ COMMAND MODE
CX→4000 (MEDIUM SPEED COUNT)
CX→1000 (FAST COUNT)

2.19. CREATING A TIME DELAY

AIM:

To create a time delay using 8086 microprocessor by the user.

APPARATUS REQUIRED: 8086 Microprocessor Kit.

ALGORITHM:

1.   Start the Program
2.   Set the DX register as 0000.
3.   Interrupt Control to AB and AC.
4.   Move the Contents of AL as 04.
5.   Interrupt to AE.
6.   Move the Contents of DL to AL.
7.   Add the AL register with 01 data.
8.   Adjust the decimal after addition.
9.   Move AL register to DL register and similarly DH register to AL.
10. Add with Carry AL and 00.
11. Move the Contents AL register to DH.
12. Move the data 4000 to CX register.
13. Jump to loop1 if zero flag is set.
14. Move the content of data 08 to AH register.
15. Compare AL and Word byte and jump to loop2.
16. Compare AL and 53 and if equal jump to loop1.
17. Jump to loop2.
18. Stop the Program.

OUTPUT:
      F – TO PAUSE
      S – TO START
      ESC – COMMAND MODE
      CX – 4000 [MEDIUM SPEED]
      CX – 1000 [FAST]

RESULT:

Thus the time delay was created using 8086 microprocessor.

CHAPTER III
8086
INTERFACING
PROGRAMS

3.1.TRAFFIC LIGHT CONTROLLER WITH 8086 INTERFACING

AIM:

To interface traffic light with 8086 microprocessor.

APPARATUS REQUIRED:

1. 8086 MP KIT
2. KEYBOARD
4. TRAFFIC LIGHT INTERFACE KIT
5. INTERFACE CABLE

ALGORITHMS:

CONDITION 1:

SOUTH TO NORTH

SOUTH TO NORTH -   VEHICLE CAN MOVE FROM

> SOUTH TO NORTH STRAIGHT, SOUTH TO EAST RIGHT,

SOUTH TO WEST LEFT, EAST TO SOUTH LEFT,

> NORTH & WEST - RED

> ALL PEDESTRIAN OFF

CONDITION 2:

ALL PEDESTRIAN ON

> NORTH, SOUTH, EAST&WEST – RED

ALL PEDSTRIAN ON

PROGRAM:

```
1000:0100  MOV    AL, 80
1000:0102  MOV    DX, 8006
1000:0105  OUT    DX, AX
START: 1000:0106 MOV AL, 61
1000:0108  MOV    DX,8000
1000:010B  OUT    DX,AL
1000:010C  MOV    AL,68
1000:010E  MOV    DX,8002
1000:0111  OUT    DX,AL
1000:0112  MOV    AL,86
1000:0114  MOV    DX,8004
1000:0117  OUT    DX,AL
1000:0118  MOV    AL,60
1000:011A  MOV    DX,8000
1000:011D  OUT    DX,AL
1000:011E  MOV    AL,48
1000:0120  MOV    DX,8002
1000:0123  OUT    DX,AL
1000:0124  MOV    AL,64
1000:0126  MOV    DX,8000
1000:0129  OUT    DX,AL
1000:012A  MOV    AL,58
1000:012C  MOV    DX,8002
1000:012F  OUT    DX,AL
1000:0130  CALL   025A
1000:0133  MOV    AL,60
1000:0135  MOV    DX,8000
1000:0138  OUT    DX,AL
1000:0139  MOV    AL,48
1000:013B  MOV    DX,8002
1000:013E  OUT    DX,AL
1000:013F  MOV    AL,62
1000:0141  MOV    DX,8000
1000:0144  OUT    DX,AL
1000:0145  CALL   0253
1000:0148  MOV    AL,60
1000:014A  MOV    DX,8000
1000:014D  OUT    DX,AL
1000:014E  MOV    AL,61
1000:0150  MOV    DX,8000
1000:0153  OUT    DX,AL
1000:0154  MOV    AL,68
1000:0156  MOV    DX,8002
1000:0159  OUT    DX,AL
1000:015A  CALL   0253
1000:015D  MOV    AL,41
1000:015F  M OV   DX,8000
1000:0162  OU T   DX,AL
1000:0163  MO V   AL,06
```

```
1000:0165 MO V    DX,8004
1000:0168 OUT    DX,AL
1000:0169 MOV    AL,49
1000:016B MOV    DX,8000
1000:016E OUT    DX,AL
1000:016F MOV    AL,26
1000:0171 MOV    DX,8004
1000:0174 OUT    DX,AL
1000:0175 CALL   025A
1000:0178 MOV    AL,41
1000:017A MOV    DX,8000
1000:017D OUT    DX,AL
1000:017E MOV    AL,06
1000:0180 MOV    DX,8004
1000:0183 OUT    DX,AL
1000:0184 MOV    AL,51
1000:0186 MOV    DX,8000
1000:0189 OUT    DX,AL
1000:018A MOV    AL,46
1000:018C MOV    DX,8004
1000:018F OUT    DX,AL
1000:0190 CALL   0253
1000:0193 MOV    AL,41
1000:0195 MOV    DX,8000
1000:0198 OUT    DX,AL
1000:0199 MOV    AL,06
1000:019B MOV    DX,8004
1000:019E OUT    DX,AL
1000:019F MOV    AL,61
1000:01A1 MOV    DX,8000
1000:01A4 OUT    DX,AL
1000:01A5 MOV    AL,86
1000:01A7 MOV    DX,8004
1000:01AA OUT    DX,AL
1000:01AB CALL   0253
1000:01AE MOV    AL,60
1000:01B0 MOV    DX,8002
1000:01B3 OUT    DX,AL
1000:01B4 MOV    AL,82
1000:01B6 MOV    DX,8004
1000:01B9 OUT    DX,AL
1000:01BA MOV    AL,62
1000:01BC MOV    DX,8002
1000:01BF OUT    DX,AL
```

FLOW CHART:

INTERFACING OF TRAFFIC LIGHT WITH 8086   CALL DELAY

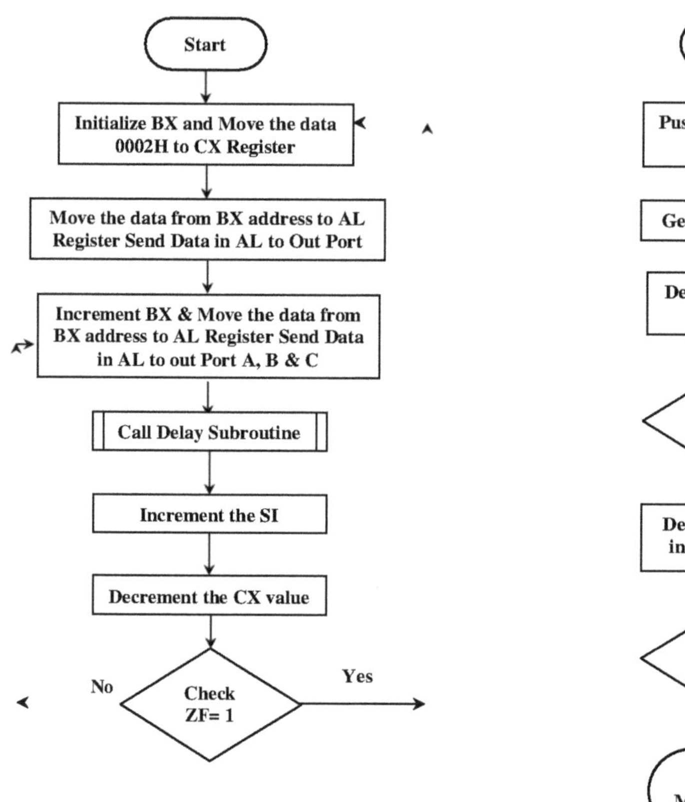

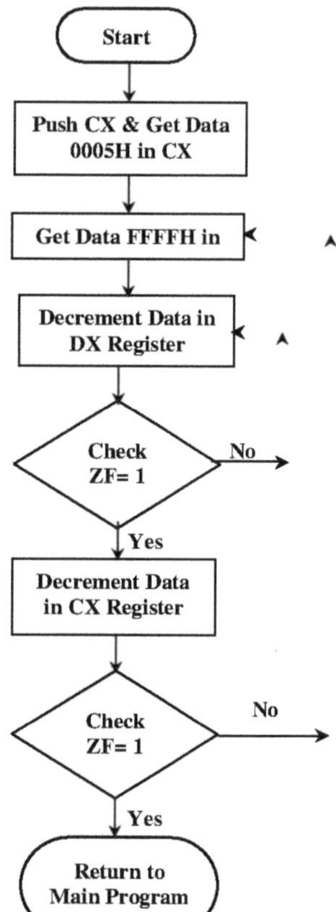

RESULT:

Thus Traffic Light was interfaced with 8086.

3.2. INTERFACING STEPPER MOTOR WITH 8086

AIM:

Thus the assembly language program to find the 1's and 2's complement of an 8-bit number using 8051 instruction set was written and executed successfully.

THEORY:

A motor in which the rotor is able to assume only discrete stationary angular position is a stepper motor. The rotary motion occurs in a step-wise manner from one equilibrium position to the next. Stepper Motors are used very wisely in position control systems like printers, disk drives, process control machine tools, etc.

The basic two-phase stepper motor consists of two pairs of stator poles. Each of the four poles has its own winding. The excitation of any one winding generates a North Pole. A South Pole gets induced at the diametrically opposite side. The rotor magnetic system has two end faces. It is a permanent magnet with one face as South Pole and the other as North Pole.

The Stepper Motor windings A1, A2, B1, B2 are cyclically excited with a DC current to run the motor in clockwise direction. By reversing the phase sequence as A1, B2, A2, B1, anticlockwise stepping can be obtained.

2-PHASE SWITCHING SCHEME:

In this scheme, any two adjacent stator windings are energized. The switching scheme is shown in the table given below. This scheme produces more torque.

ANTICLOCKWISE					CLOCKWISE				
STEP	A1	A2	B1	B2	DATA	STEP	A1	A2	B1	B2	DATA
1	1	0	0	1	9h	1	1	0	1	0	Ah
2	0	1	0	1	5h	2	0	1	1	0	6h
3	0	1	1	0	6h	3	0	1	0	1	5h
4	1	0	1	0	Ah	4	1	0	0	1	9h

ADDRESS DECODING LOGIC:

The 74138 chip is used for generating the address decoding logic to generate the device select pulses; CS1 & CS2 for selecting the IC 74175.The 74175 latches the data bus to the stepper motor driving circuitry.

Stepper Motor requires logic signals of relatively high power. Therefore, the interface circuitry that generates the driving pulses uses silicon Darlington pair transistors. The inputs for the interface circuit are TTL pulses generated under software control using the Microcontroller Kit. The TTL levels of pulse sequence from the data bus are translated to high voltage output pulses using a buffer 7407 with open collector.

PROCEDURE:

Enter the above program starting from location 1000 and execute the same. The stepper motor rotates. Varying the count at DX register pair can vary the speed. Entering the data in the look-up TABLE in the reverse order can vary direction of rotation.

APPARATUS REQUIRED:
1. 8086 MP kit
2. Keyboard
4. Stepper motor interface kit
5. Interface cable

ALGORITHM:

For running stepper motor in clockwise and anticlockwise directions.
1. Get the first data from the lookup table
2. Initialize counter and move data into accumulator
3. Drive the stepper motor circuitry and introduce delay
4. Decrement the counter
5. If it is not zero repeat from step 3
6. Repeat the above procedure for both backward and forward directions.
7. Stop the program

PROGRAM:

```
        MOV CX, 8000
        MOV AL, 80
        MOV DX, 8006
        OUT DX, AL
        MOV AL, 00
        MOV DX, 8000
        OUT DX, AL
AGAIN: MOV AL, 06H/03H
        OUT DX, AL
        INT AAH
        MOV AL, 0CH
        OUT DX, AL
        INT AAH
        MOV AL, 09H/0CH
        OUT DX, AL
        INT AAH
        MOV AL, 03H
        OUT DX, AL
        INT AAH
        JMP AGAIN
        INT A5
```

CLOCKWISE DIRECTION

06 0C 09 03

ANTICLOCKWISE DIRECTION

03 09 0C 06

RESULT:

Thus a stepper motor was interfaced with 8086 and run in forward and reverse directions at various speeds.

3.3 .DIGITAL CLOCK IN REAL TIME

AIM:

To write an 8086 ALP to display a digital clock in real time.

APPARATUS REQUIRED: 8086 Microprocessor kit

ALGORITHM:

1. Start the program.
2. Load the contents of AX to EX Segment.
3. Create delay by moving contents to consecutive memory locations.
4. Send through output port address 8c07 and 8c03.
5. Jump to display the clock.
6. Set the program.

PROGRAM:

```
                MOV AX,0000
                NOP
                MOV ES,AX
                INT    AC
                MOV AL,00
                MOV  [1000], AL
                MOV  [1001], AL
                MOV  [1002], AL
                MOV  [1003], AL
                MOV [1004], AL
                PUSH CS
                POP    CX
                MOV SI, ISR
                MOV AL, 02
                INT    BE
                MOV AL,7E
                MOV DX,8C07
                OUT   DX, AL
                MOV AL, 24
                MOV DX, 8C03
                OUT   DX, AL
                MOV AL, F4
                MOV DX, 8C03
                OUT   DX, AL
                JMP    DSPLY

LOOP1 (0136) : MOV AL, [1003]
                CMP   AL, 00
                JZ       LOOP1
                MOV   AL, [1002]
                ADD   AL, 01
                DAA
                CMP   AL, 60
                JZ       NXT1
                MOV   [1002], AL
                MOV   AL, 00
                MOV   [1003], AL
                JMP    DSPLY

NXT1 (0151) : MOV AL, 00
                MOV   [1002], AL
                MOV   AL, [1001]
                ADD   AL, 01
                DAA
                CMP   AL, 60
                JZ       NXT2
```

```
                MOV  [1001], AL
                MOV  AL, 00
                MOV  [1003], AL
                JMP    DSPLY

NXT2 (016A):  MOV AL, 00
                MOV  [1001], AL
                MOV  AL, [1000]
                ADD  AL, 01
                DAA
                CMP  AL, 13
                JZ    END
                MOV  [1000], AL
                MOV  AL, 00
                MOV  [1003], AL

DSPLY (0181):  MOV DL, 20
                MOV  AH, 02
                INT   A2
                MOV  AL, 02
                MOV  DL, [1000]
                INT   AE
                MOV  DL, 20
                MOV  AH, 02
                INT   A2
                MOV  AL,02
                MOV  DL, [1001]
                INT   AE
                MOV  DL, 20
                MOV  AH, 02
                INT   A2
                MOV  AL,02
                MOV  DL, [1002]
                INT   AE
                INT   AB
                MOV  AH, 0B
                INT AB
                MOV AH, 0B
                INT   A1
                CMP  AL,00
                JNZ   END
                MOV BX, LOOP1
                JMP   BX

END (01BA): MOV AL, 76
                MOV DX, 8C07
                OUT   DX, AL
                INT   A4
                NOP
```

```
ISR (01C3):  PUSH  AX
             MOV  AL, [1004]
             INC AL
             CMP  AL,  28
             JNZ    SKIP1
             MOV AL, 00
             MOV [1004], AL
             MOV AL, 01
             MOV [1003], AL
             JMP    SKIP3

SKIP1 (01D9): MOV [1004], AL
SKIP3 (01DC): POP AX
             RET

             MOV AX, 0000
             NOP
             MOV ES, AX
             INT    AC
             MOV AL, 00
             MOV   ]1000], AL
             MOV   [1001], AL
             MOV   [1002], AL
             MOV   [1003], AL
             MOV [1004], AL
             PUSH CS
             POP    CX
             MOV SI, ISR
             MOV AL, 02
             INT    BE
             MOV   AL, 7E
             MOV DX, 8C07
             OUT   DX, AL
             MOV AL, 24
             MOV DX, 8C03
             OUT   DX, AL
             MOV AL, F4
             MOV DX, 8C03
             OUT   DX, AL
             JMP    DSPLY

LOOP1 (0136) : MOV AL,[1003]
             CMP    AL, 00
             JZ     LOOP1
             MOV    AL, [1002]
             ADD    AL, 01
             DAA
             CMP    AL, 60
             JZ     NXT1
             MOV    [1002], AL
             MOV    AL, 00
```

```
                MOV  [1003], AL
                JMP  DSPLY

NXT1 (0151): MOV AL, 00
                MOV  [1002], AL
                MOV  AL, [1001]
                ADD  AL, 01
                DAA
                CMP  AL, 60
                JZ   NXT2
                MOV  [1001], AL
                MOV  AL, 00
                MOV  [1003], AL
                JMP  DSPLY

NXT2 (016A):  MOV AL, 00
                MOV  [1001], AL
                MOV  AL, [1000]
                ADD  AL, 01
                DAA
                CMP  AL, 13
                JZ   END
                MOV  [1000], AL
                MOV  AL, 00
                MOV  [1003], AL

DSPLY (0181): MOV DL, 20
                MOV  AH, 02
                INT  A2
                MOV  AL,02
                MOV  DL, [1000]
                INT  AE
                MOV  DL, 20
                MOV  AH, 02
                INT  A2
                MOV  AL, 02
                MOV  DL, [1001]
                INT  AE
                MOV  DL, 20
                MOV  AH, 02
                INT  A2
                MOV  AL, 02
                MOV  DL, [1002]
                INT  AE
                INT  AB
                MOV  AH, 0B
                INT  A1
                CMP  AL,00
                JNZ  END
                MOV  BX, LOOP1
                JMP  BX
```

END (01BA): MOV AL, 76
            MOV DX, 8C07
            OUT   DX, AL
            INT   A4
            NOP

ISR (01C3):   PUSH AX
            MOV AL, [1004]
            INC AL
            CMP AL, 28
            JNZ   SKIP1
            MOV AL, 00
            MOV [1004], AL
            MOV AL, 01
            MOV [1003], AL
            JMP   SKIP3

SKIP1 (01D9): MOV [1004], AL
SKIP3 (01DC): POP AX
            RET

RESULT:

Thus the assembly language program to display a digital clock in real time was written and executed successfully.

3.5 INTERFACING 8279 KEYBOARD AND DISPLAY CONTROLLER WITH 8086

AIM:

To interface 8279 Programmable Keyboard Display Controller to 8086 Microprocessor.

APPARATUS REQUIRED:
1. 8086 Microprocessor tool kit.
2. 8279 Interface board.
3. VXT parallel bus.

ALGORITHM:

1. Start the program.
2. Initialize the counter.
3. Set 8279 for 8 digit character display right entry.
4. Set 8279 for clearing the display.
5. Write the command to display.
6. Load the accumulator with character and display it.
7. Introduce the delay.
8. Repeat the above steps.

PROGRAM:

```
START: MOV AL, 18
MOV DX, 01E2
OUT DX, AL
MOV AL, DF
OUT DX, AL
MOV AL, 8F
OUT DX, AL
MOV CL, 10
CALL DISPLAY
JMP START
DISPLAY: MOV BX, 1100
NEXT: MOV AL, [BX]
MOV DX, 01E0
OUT DX, AL
INC BX
DEC CL
JNZ NEXT
INT A5
```

Connections:
1. Connect EN/U8 tag to ground tag to enable U8 decoder.
2. Connect EN/DSP tag to 8th output tag of the U8 decoder.

INPUT

$1100 = C0_H$	$1108 = 80_H$
$1101 = F9_H$	$1109 = 90_H$
$1102 = A4_H$	$110A = 88_H$
$1103 = B0_H$	$110B = 83_H$
$1104 = 99_H$	$110C = C6_H$
$1105 = 92_H$	$110D = A1_H$
$1106 = 82_H$	$110E = 86_H$
$1107 = F8_H$	$110F = 8E_H$

DISPLAY	DISPLAY SEGMENTS								HEX VALUE (input)
	dp	g	f	e	d	c	b	a	
0	1	1	0	0	0	0	0	0	C0
1	1	1	1	1	1	0	0	1	F9
2	1	0	1	0	0	1	0	0	A4
3	1	0	1	1	0	0	0	0	B0
4	1	0	0	1	1	0	0	1	99
5	1	0	0	1	0	0	1	0	92
6	1	0	0	0	0	0	1	0	82
7	1	1	0	1	1	0	0	0	D8
8	1	0	0	0	0	0	0	0	80
9	1	0	0	1	0	0	0	0	90

SEVEN SEGMENT DISPLAY

```
      a
     ___
  f |   | b
     ___
      g
  e |   | c
     ___  . dp
      d
```

COMMON ANODE CONFIGURATION
Segment glow = 0
Segment not glow = 1

OUTPUT:

Step	Data bus	Display
Start	10	-
1	DF	-
2	90	BLANK
3	C0	BLANK
4	F9	0
5	A4	1
6	B0	2
7	99	3
8	92	4
9	82	5
10	F8	6
11	80	7
12	90	8
13	88	9
14	83	A
15	C6	B
16	A1	C
17	86	D
18	8E	E
19	10	F

RESULT:

Thus 8279 controller was interfaced with 8086 and program for rolling display was executed successfully.

3.5 INTERFACING ADC WITH 8086

AIM:

To interface ADC with 8086.

APPARATUS REQUIRED:

1. 8086 Microprocessor tool kit.
2. VBMB-003 Interface board.
3. VXT parallel bus.
4. Multimeter.

ALGORITHM:

Select the channel and latch the address.
Send the start conversion pulse.
Read EOC signal.
If EOC=1 continue, else go to before step.
Read the digital output.
Store it in a Memory location.

Calculation procedure:

$$\text{DIGITAL OUTPUT} = \frac{\text{ANALOG INPUT X 255}}{5}$$

Sl.No	Analog Voltage(V)	Digital output (HEX)

PROGRAM:

```
                MOV DX, 8807
                MOV AL, 81
                OUT DX, AL
                MOV DX, 8803
                MOV AL, 01
                OUT DX, AL
                MOV DX, 8807
                MOV AL, 09
                OUT DX, AL
                MOV AL, 08
                OUT DX, AL
                MOV AL, 83
                OUT DX, AL
                INT AC
                MOV DX, 8807
                MOV AL, 0D
                OUT DX, AL
                MOV AL, OC
                OUT DX, AL
                MOV DX, 8805
LOOP1 :         IN AL, DX
                AND AL, 02
                JNZ LOOP1
LOOP2:          IN AL, DX
                AND AL, 02
                JZ LOOP2
                MOV AL, 0B
                MOV DX, 8807
                OUT DX, AL
                MOV DX, 8803
                IN AL, DX
                MOV [1100], AL
                INT A5
```

RESULT:

Thus the ADC was interfaced with 8086 and Analog to Digital conversion
was achieved.

3.6 INTERFACING DAC WITH 8086

AIM:

To generate a saw tooth wave, triangular and square-wave at the output of DAC.

APPARATUS REQUIRED:

1. 8086 Microprocessor tool kit.
2. VBMB-002 Interface board.
3. VXT parallel bus.
4. Power supply.
5. CRO.

ALGORITHM:

Saw tooth wave
1. Start the Program
2. Make the accumulator '00'.
3. Display accumulator content at port.
4. Increment accumulator by 1.
5. If it is not equal to '0' go to L1.
6. Jump to start unconditionally for continuous waveform.

Triangular wave
1. Start the Program
2. Make the accumulator '00'.
3. Move the contents of AL register.
4. Display contents of accumulator content at port.
5. Increment accumulator by 1.
6. If AL≠0, jump to loop1.
7. Set accumulator as FF=maximum.
8. Display accumulator content as port.
9. Decrement accumulator by 1.
10. Jump to start unconditionally for continuous waveform.

Square wave
1. Start the program.
2. Move 0 to accumulator.
3. Display 0 at port.
4. Call the delay subroutine.
5. Move FF into accumulator.
6. Display at port.
7. Call the delay subroutine.
8. Jump to Start.
9. Move 05 to register BL to give delay.
10. Move FF to register CL and decrement CL.
11. Return to Main program.

PROGRAM:

To create a saw tooth waveform at DAC1 in Micro-86 LCD kit

```
START: MOV AL, 80
       MOV DX, 8807
       OUT DX, AL
       MOV DX, 8801
       MOV AL, 00
  L1:  OUT DX, AL
       INC AL
       JNZ L1
       JMP START
       RET
```

To create a triangular waveform at DAC1 in Micro-86 LCD kit

```
START:   MOV AL, 80
         MOV DX, 8807
         OUT DX, AL
         MOV DX, 8801
         MOV AL, 00
  L1:    OUT DX, AL
         INC AL
         JNZ L1
         MOV AL, FF
  L2:    OUT DX, AL
         DEC AL
         JNZ  L2
         JMP START
         RET
```

To create a square waveform at DAC1 in Micro-86 LCD kit

```
START:   MOV AL, 80
         MOV DX, 8807
         OUT DX, AL
         MOV DX, 8801
         MOV AL, FF
         OUT DX, AL
         CALL DELAY
         MOV DX, 8801
         MOV AL, 00
         OUT DX, AL
         CALL DELAY
```

```
        JMP START
DELAY:   MOV BL, 05
    L1: MOV CL, FF
    L2: DEC CL
        JNZ L2
        DEC BL
        JNZ L1
        RET
```

RESULT:

Thus VBMB-002 was interfaced with 8086 and various waveforms are generated at the output of DAC.

3.7 PARALLEL COMMUNICATION INTERFACE

AIM:

To establish parallel communication between two MP kits using 8255.

APPARATUS REQUIRED:

1. 8086 Microprocessor tool kit.
2. VXT parallel bus.
3. 8255 Interface board

ALGORITHM:

1. Initialize accumulator to hold Control word.
2. Store control word in control word register.
3. Read data port A.
4. Store data from port A in memory.
5. Place contents in port B.
6. Stop the program.

PROGRAM:

```
MOV AL, 90
MOV DX, 01E6
OUT DX, AL
MOV DX, 01E0
IN AL, DX
MOV DX, 01E2
OUT DX, AL
MOV [1100], AL
INT A5
```

OBSERVATION:

INPUT AT PORT A:

PA7	PA6	PA5	PA4	PA3	PA2	PA1	PA0	
0	0	0	0	1	1	1	1	$=0F_H$

OUTPUT AT PORT B:

PB7	PB6	PB5	PB4	PB3	PB2	PB1	PB0	
0	0	0	0	1	1	1	1	$=0F_H$

$1100 = 0F_H$

RESULT:

Thus the 8255 has been interfaced to 8086 microprocessor and parallel communication is established.

3.8 SERIAL COMMUNICATION INTERFACE

AIM:

To establish serial communication between two MP kits using 8251.

APPARATUS REQUIRED:
1. 8086 Microprocessor tool kit.
2. RS 232C cable.
3. 8251 Interface board

ALGORITHM:

1. Initialize 8086 and 8251 to check the Transmission and Reception of a Character.
2. Initialize 8086 to give an output.
3. The command word and mode word is written to the 8251 for subsequent operations.
4. The status word is read from the 8251 on completion of a serial I/O operation, or when the host CPU is checking the status of the device before starting the next operation.
5. Stop the program.

PROGRAM:

```
            MOV SI, 1100
            MOV DI, 1200
            MOV AL, 00
            MOV DX, 01E2
            OUT DX, AL
            INT AC
            OUT DX, AL
            INT AC
            OUT DX, AL
            INT AC
            MOV AL, 40
            OUT DX, AL
            INT AC
            MOV AL, 4D
            OUT DX, AL
            INT AC
            MOV AL, 27
            OUT DX, AL
LOOP1 : IN AL, DX
            AND AL, 01
```

```
                JZ LOOP1
                MOV AL, [SI]
                MOV DX, 01E0
                OUT DX, AL
                MOV DX, 01E2
                IN AL, DX
                AND AL, 02
                CMP AL, 02
                JNZ LOOP2
                MOV DX, 01E0
                OUT DX, AL
        LOOP 2: MOV DX, 01E2
                IN AL, DX
                AND AL, 02
                CMP AL, 02
                JNZ LOOP2
                MOV DX, 01E0
                IN AL, DX
                MOV [DI], AL
                INT A5
```

OBSERVATION:

INPUT:
1100   AA

OUTPUT:

Step	Data bus	RxD	TxD	comments
Start	00	1	1	DUMMY CODE
1	00	1	1	DUMMY CODE
2	00	1	1	DUMMY CODE
3	40	1	1	RESET code for 8251
4	4D	1	1	MODE WORD, 8 BIT NO, NO OF STOP BITS=1, BAUD RATE FACTOR=1.
5	27	1	1	COMMAND WORD, RTS OUTPUT=0, ENABLE RECEIVER, TRANSMITTER.
6	85	1	1	READS STATUS OF BIT "TxRDY"
7	AA	1	1	I/P DATA ON THE DATA BUS OF 8251
8	81	1	1	DATA IS READY IN TRANSMITTER BUFFER TO BE TRANSMITTED
9	81	0	0	CHECKING THE STATUS OF TxD PIN FOR "ONE" BEFORE SENDING THE START BIT.

10	81	0	0	THE DATA IS SERIALLYBEING TRANSMITTED AND RECEIVED THROUGH SERIAL PORT, START BIT=0
11	81	0	0	D0=0
12	81	1	1	D1=1
13	81	0	0	D2=0
14	81	1	1	D3=1
15	81	0	0	D4=0
16	81	1	1	D5=1
17	81	0	0	D6=0
18	81	1	1	D7=1
19	87	1	1	STOP BIT=1 DATA IS AVAILABLE ON 8251 WHICH IS TO BE READ
20	AA	1	1	DATA IS BEING READ FROM RECIEVER BUFFER

STEP 21... control will automatically go to command mode and check the data in mem location

RESULT:

Thus the 8251 has been interfaced to 8086 microprocessor and the serial communication between two microprocessor kits has been studied.

CHAPTER IV
PROGRAMMING WITH 8051

4.1 – BIT ADDITION AND SUBTRACTION

PROGRAM:

ADDITION/SUBTRACTION

```
MOV DPTR, #6200
MOV R2, #00
MOVX A, @DPTR
MOV R1, A
INC DPTR
MOVX A, @DPTR
ADD A, R1/SUBB A, R1
JNC
INC DPTR
MOVX @DPTR, A
MOV A, R2
INC DPTR
MOVX @DPTR, A
RET
```

4.1 – BIT ADDITION AND SUBTRACTION

AIM:

To write an assembly language program to add and subtract the two 8-bit numbers using microcontroller instruction set.

APPARATUE REQUIRED: 8051 Microprocessor kit

ALGORITHM:

1. Load Accumulator A with any desired 8 bit data.
2. Load the register R1 with the second 8 bit data.
3. Add and subtract two 8 bit numbers.
4. Store the result.
5. Stop the Program.

OBSERVATION:

ADDITION: WITHOUT CARRY

Input Data:	Output Data:
6200 – 25	6202 – 4B
6200 – 26	6203 – 00

ADDITION: WITH CARRY

Input Data:	Output Data:
6200 – EE	6202 – B9
6201 – CB	6203 – 01

SUB: WITHOUT BORROW

Input Data:	Output Data:
6200 – 65	6202 – 4D
6201 – 18	6203 – 00

SUB: WITHOUT BORROW

Input Data:	Output Data:
6200 – 01	6202 – 02
6201 – FF	6203 – FF

RESULT:

Thus, the assembly language program to add and sub the two 8-bit numbers using 8051 instruction set was written and executed successfully.

PROGRAM:

```
MOV DPTR, #6100
MOVX A, @DPTR
MOV F0, A
INC DPTR
MOVX A, @DPTR
MUL AB/DIV AB
INC DPTR
MOVX @DPTR, A
MOV A, F0
INC DPTR
MOVX @DPTR, A
RET
```

<u>4.2– BIT MULTIPLICATION AND DIVISION</u>

AIM:

To write an assembly language program to multiply and divide the two 8-bit numbers using microcontroller instruction set.

APPARATUS REQUIRED: 8051 Microprocessor kit

ALGORITHM:

1. Clear C-register for carry.
2. Move the first data to Accumulator.
3. Move the second data to B-register.
4. Multiply the second data with Accumulator.
5. The higher order of the result is in B-register.
6. The lower order of the result is in Accumulator for division quotient is stored in A and remainder in B register.
7. Store the sum in memory pointed by DPTR.

OBSERVATION:

MULTIPLICATION:

Input Data:	**Output Data:**
6100 – 06	6202 – 12
6101 – 03	

DIVISION:

Input Data:	**Output Data:**
6100 – 02	6102 – 04
6101 – 08	

RESULT:

Thus, the assembly language program to multiply and divide the two 8-bit numbers using 8051 instruction set was written and executed.

PROGRAM:

AND/EX-OR/OR
```
MOV DPTR, #6100
MOVX A, @DPTR
MOV R1, A
INC DPTR
MOVX A, @DPTR
ANL A, R1/XRL A, R1/ORL A, R1
INC DPTR
MOVX @DPTR, A
RET
```

NOT
```
MOV DPTR, #6100
MOVX A, @DPTR
CPL A
INC DPTR
MOVX @DPTR, A
RET
```

<h1 align="center">4.3. LOGICAL OPERATIONS</h1>

AIM:

To write an assembly language program to perform logical operations on 8-bit data using 8051 microcontroller.

APPARATUS REQUIRED: 8051 µc kit, keyboard board, power supply

ALGORITHM:

1. Load 8 bit data in accumulator
2. Load R1 register with second 8 bit data.
3. Perform AND/OR/XOR operation on two data's.
4. Complement the Result.
5. Store the result in memory locations.
6. Stop the program.

OBSERVATION:

AND

Input Data:	Output Data:
6100 – BC	6102 – AC
6101 - AE	6103 – 00

OR

Input Data:	Output Data:
6100 – BC	6102 – BE
6101 - AE	6103 – 00

EX - OR

Input Data:	Output Data:
6100 – 08	6102 – 07
6101 – 0F	

NOT

Input Data:	Output Data:
6100 – 08	6102 – F7

RESULT:

Thus, the assembly language program was performed for logical operations on 8-bit data.

4.4 SQUARE OF AN 8 BIT NUMBER

PROGRAM:

SQUARE OF AN 8-BIT NUMBER:

```
MOV DPTR, #6100
MOVX A, @DPTR
MOV F0, A
MUL AB
INC DPTR
MOVX @DPTR, A
INC DPTR
MOV A, F0
MOVX @DPTR,A
RET
```

4.4 SQUARE OF AN 8 BIT NUMBER

AIM:

To write an assembly language program to find square and cube of an 8 bit number using 8051 microcontroller.

APPARATUS REQUIRED: 8051 µc kit, keyboard board, power supply

ALGORITHM:

1. Load accumulator A with 8 bit data.
2. Move the 8 bit data to B register.
3. Multiply A and B to obtain square of a number.
4. Move the result to B register.
5. Move to the memory locations.
6. Stop the program.

OBSERVATION:

Input Data:	**Output Data:**
6100 - 02	6101 – 04 (SQUARE)

RESULT:

Thus the square of an 8-bit number was performed using 8051 microcontroller.

4.5 CUBE OF AN 8 BIT NUMBER

PROGRAM:

CUBE OF AN 8-BIT NUMBER:

```
MOV DPTR, #6100
MOVX A, @DPTR
MOV R0, A
MOV F0, A
MUL AB
PUSH F0
MOV F0, A
MOV A, R0
MUL AB
INC DPTR
MOVX @DPTR, A
MOV A, F0
MOV R1, A
POP F0
MOV A, R0
MUL AB
ADD A, R1
INC DPTR
MOVX @DPTR, A
MOV A, F0
ADDC A, #00
INC  DPTR
MOVX @DPTR, A
RET
```

4.5 CUBE OF AN 8 BIT NUMBER

AIM:

To write an assembly language program to find cube of an 8 bit number using 8051 microcontroller.

APPARATUS REQUIRED: 8051 μc kit, keyboard board, power supply

ALGORITHM:

1. Load accumulator A with 8 bit data.
2. Move the 8 bit data to B register.
3. Multiply A and B to obtain Cube of a Number.
4. Move to the memory locations.
5. Stop the program.

OBSERVATION:

Input Data:	**Output Data:**
6100 - 02	6102 – 08 (CUBE)

RESULT:

Thus the cube of an 8-bit number was performed using 8051 microcontroller

4.6 ONE'S AND TWO'S COMPLEMENT

PROGRAM:

```
MOV DPTR, #6100
MOVX A, @DPTR
CPL A
INC  DPTR
MOVX @DPTR, A
INC A
INC  DPTR
MOVX @DPTR, A
RET
```

4.6 ONE'S AND TWO'S COMPLEMENT

AIM:

To write an assembly language program to find the 1's and 2's complement of an   8-bit number using microcontroller instruction set.

APPARATUS REQUIRED: 8051 µc kit, keyboard board, power supply

ALGORITHM:

1. Move the data to Accumulator.
2. Complement the accumulator.
3. Move the one's complement output to the memory 6500H.
4. Add 01H with accumulator.
5. Move the two's complement output to the memory 6501H.

OBSERVATION:

Input Data	**Output Data:**
6000 - 03	6100 – FC (1's Complement)
	6102 – FD (2's Complement)

RESULT:

Thus the assembly language program to find the 1's and 2's complement of an 8-bit number using 8051 instruction set was written and executed successfully.

4.7 CODE CONVERSION

<u>UNPACKED BCD TO ASCII</u>

PROGRAM:

```
MOV DPTR, #6100
MOVX A, @DPTR
MOV F0, A
ANL A, #0F
ADD A, #30
INC DPTR
MOVX @DPTR, A
MOV A, F0
ANL A, #F0
SWAP A
ADD A, #30
INC DPTR
MOVX @DPTR, A
RET
```

UNPACKED BCD TO ASCII

AIM:

To write an assembly language program to convert unpacked BCD to ASCII of an 8-bit number using 8051 microcontroller.

ALGORITHM:

1. Load a 8 bit data into accumulator.
2. Move contents of A to R1 and Clear A register.
3. Add contents of A AND R1.
4. Increment DPTR register.
5. Add A and R1 contents.
6. Jump to loop if carry flag is set.
7. Add A contents and 07.
8. Increment DPTR register.
9. Stop the Program.

OBSERVATION:

Input Data:	Output Data:
6100 – 46	6102 – 36
	6101 – 34

RESULT:

Thus the assembly language program to convert unpacked BCD to ASCII of an 8-bit number using 8051 microcontroller was written and executed successfully.

4.8 INTERFACING SEVEN SEGMENT DISPLAY WITH 8051

AIM:

To interface seven segment display with 8051 to display hexadecimal values from 0 to 9 for common anode configuration.

APPARATUS REQUIRED: 8051 development kit, ISP programmer, Keil μVision3 compiler, ribbon cables, 5V power supply.

INTERFACING DIAGRAM:

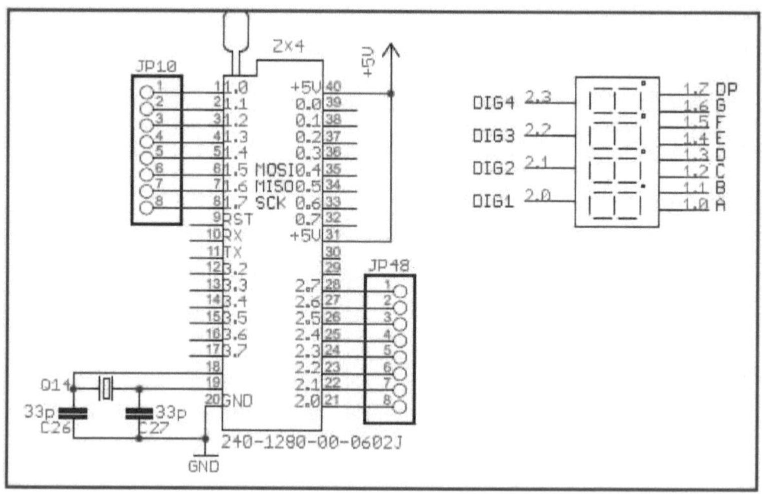

PROGRAM:

```c
#include<reg51.h>        // include the header file
void delay_1s()          // Function for creating delay in milliseconds.
{
int d;
for(d=0;d<=20;d++)
{
TMOD=0x01;
TL0=0xFD;
TH0=0x04B;
TR0=1;                   // start timer.
```

```c
while(TF0==0);              // run until TF turns to 1
TR0=0;                      // stop timer
TF0=0;                      // reset the flag
}
}

void main()
{
    unsigned char
no_code[]={0x3F,0x06,0x5B,0x4F,0x66,0x6D,0x7D,0x07,0x7F,0x67}; //Array
for hex values (0-9) for common anode 7 segment
    int k;
    while(1)                // repeat the loop infinitely
    {
        for(k=0;k<10;k++)
        {
         P2=no_code[k];
         delay_1s();
        }
    }
}
```

RESULT

Thus the interfacing seven segment display with 8051 to display hex values was performed.

4.9 INTERFACING LCD WITH 8051

AIM:

To interface LCD with 8051 to display characters using embedded C programming.

APPARATUS REQUIRED: 8051 development kit, ISP programmer, Keil μVision3 compiler, ribbon cables, 5V power supply.

INTERFACING DIAGRAM:

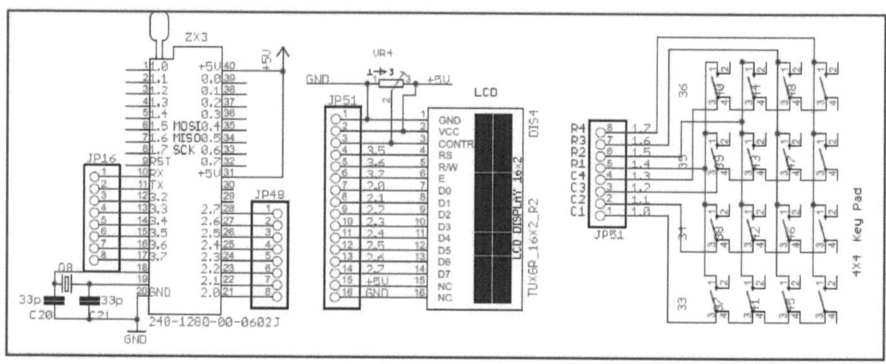

PROGRAM:

```c
include<reg51.h>                //include the header file
#define LCD_PORT P2
sbit rs=P3^5;                   //defining name for the pin
sbit en=P3^7;
sbit D7=P2^7;
sbit rw=P3^6;

void busy();
void CMD_WRT(unsigned char); // function definitions
void LCD_WRT(unsigned char *);
void DATA_WRT(unsigned char);
void DELAY();
void main()
{
unsigned char CMD[]={0x38,0x01,0x0f,0x06,0x80},
                    TEMP1,i;            //array declaration
   for(i=0;i<5;i++)                     //creating a loop
  {
  TEMP1=CMD[i];
       CMD_WRT(TEMP1);                  // writing command to LCD
  }
     DELAY();
```

```c
            CMD_WRT(0X01);
            CMD_WRT(0X80);
            LCD_WRT("16X2 LCD DISPLAY");
            DELAY();
                DELAY();
                while(1); }                          // indefinite loop
    void DELAY()                                     // calling delay function
    {
            unsigned int X=6000000;
            while(X--); }
    void busy()                                      //checking the status of LCD
    {
            D7=1;
            rs=0;
            rw=1;
            while(D7!=0)
            {
                    en=0;
                    en=1;
            }
    }
        void CMD_WRT(unsigned char val)// passing command to LCD
        {
        busy();
        LCD_PORT=val;
        rs=0;
        rw=0;
        en=1;
        en=0;
        }
        void LCD_WRT(unsigned char *string) //writing string to LCD
        {
        while(*string)
        DATA_WRT(*string++);
        }
        void DATA_WRT(unsigned char ch) // writing data
        {
        busy();
        LCD_PORT = ch;
        rs=1;
        rw=0;
        en=1;
        en=0;}
```

RESULT

Thus the interfacing LCD with 8051 to display characters was performed.

4.10 INTERFACING DC MOTOR WITH 8051

AIM:

To interface DC motor with 8051 and make it rotate in clockwise and anticlockwise direction using embedded C programming.

APPARATUS REQUIRED: 8051 development kit, ISP programmer, Keil µVision3 compiler, ribbon cables, 5V power supply.

INTERFACING DIAGRAM:

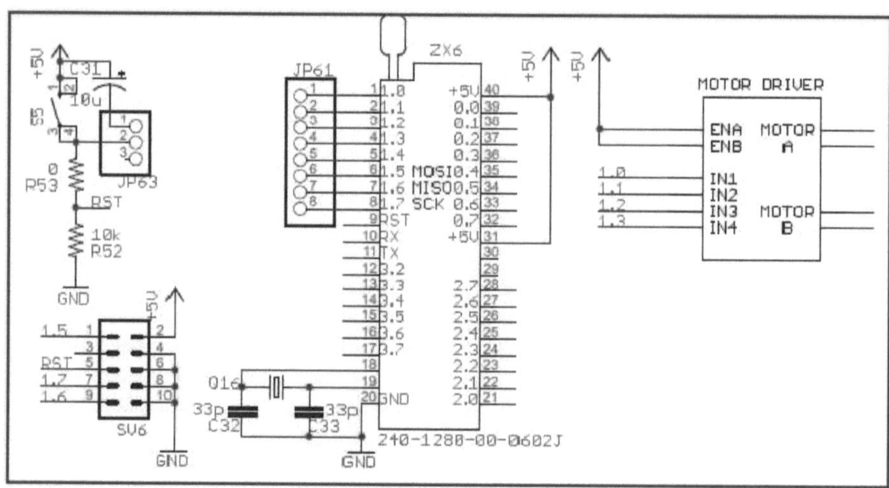

PROGRAM:

```
#include<reg51.h>              //include header file
sbit switch1=P2^0;             //initialize switch1 to P2.0
sbit switch2=P2^1;             //initialize switch2 to P2.1
sbit clk=P3^0;
sbit anticlk=P3^1;

void main()
{

        switch1=switch2=1;        //making P2.0 and P2.1 as inputs
        switch1=switch2=0;
        clk=anticlk=0;
        while(1)                  // repeat the loop infinitely
        {
            if((switch1))
                clk=1;
            else if((switch2))
```

```
                anticlk=1;
        else
                P3=0x00;


    }
}
```


RESULT

Thus the interfacing the DC motor with 8051 to make it rotate in clockwise and anticlockwise direction was performed.

PROGRAM OUTCOMES:

At the end of the course, the students will be able to
- Write embedded C program for 8051 microcontroller
- Build a complete microcontroller based system
- Debug programming errors
- Develop microcontroller automotive applications

ALGORITHM FOR EXECUTING PROGRAMS IN CONTENT BEYOND SYLLABUS
1. Design the circuit
2. Simulate the circuit using Proteus software
3. Compile the embedded C code using Keil software
4. Create the hex file
5. Download the hex file into the chip using PROGisp
6. View the desired output in the kit

CHAPTER V

5.1 8086 VIVA QUESTIONS

1. Give examples for 8 / 16 / 32 bit Microprocessor?
 8-bit Processor - 8085 / Z80 /6800; 16-bit Processor - 8086 / 68000 /
Z8000; 32-bit Processor - 80386 / 80486.

2. What is 1st / 2nd / 3rd / 4th generation processor?
 The processor made of PMOS/ NMOS / HMOS / HCMOS technology is called
1st / 2nd / 3rd / 4th generation processor, and it is made up of 4 / 8 / 16 /
32 bits.

3. What is the difference between microprocessor and microcontroller?
In Microprocessor more op-codes, few bit handling instructions. But in
Microcontroller: fewer op-codes, more bit handling Instructions, and also it is
defined as a device that includes micro processor, memory, & input / output
signal lines on a single chip.

4. What is meant by LATCH?
Latch is a D- type flip-flop used as a temporary storage device controlled by a
timing signal, which can store 0 or 1. The primary function of a Latch is data
storage. It is used in output devices such as LED, to hold the data for display.

5. What is the difference between primary & secondary storage device?
In primary storage device the storage capacity is limited. It has a volatile
memory. In secondary storage device the storage capacity is larger. It is a
nonvolatile memory. Primary devices are: RAM / ROM. Secondary devices are:
Floppy disc / Hard disk.

6. Difference between static and dynamic RAM?
Static RAM: No refreshing, 6 to 8 MOS transistors are required to form one
memory cell, Information stored as voltage level in a flip flop. Dynamic RAM:
Refreshed periodically, 3 to 4 transistors are required to form one memory cell,
Information is stored as a charge in the gate to substrate capacitance.
7. What is interrupt?
Interrupt is a signal send by external device to the processor so as to request
the processor to perform a particular work.

8. What is cache memory?
Cache memory is a small high-speed memory. It is used for temporary storage
of data & information between the main memory and the CPU (center
processing unit). The cache memory is only in RAM.

9. What is called .Scratch pad of computer?
Cache Memory is scratch pad of computer.

10. Differentiate between RAM and ROM?
RAM: Read / Write memory, High Speed, Volatile Memory. ROM: Read only memory, Low Speed, Non-Volatile Memory.

11. What is a compiler?
Compiler is used to translate the high-level language program into machine code at a time. It doesn't require special instruction to store in a memory, it stores automatically. The Execution time is less compared to Interpreter.

12. Which processor structure is pipelined?
All x86 processors have pipelined structure.

13. What is flag?
Flag is a flip-flop used to store the information about the status of processor and the status of the instruction executed most recently

14. What is stack?
Stack is a portion of RAM used for saving the content of Program Counter and general purpose registers.

15. Can ROM be used as stack?
ROM cannot be used as stack because it is not possible to write to ROM.

16. What is NV-RAM?
Nonvolatile Read Write Memory also called Flash memory. It is also know as shadow RAM

17. What are the flags in 8086?
In 8086 carry flag, Parity flag, Auxiliary carry flag, Zero flag, Overflow flag, Trace flag, Interrupt flag, Direction flag, and Sign flag.

18. What are the various interrupts in 8086?
Maskable interrupts Non-Maskable interrupts.

19. What is meant by Maskable interrupts?
An interrupt that can be turned off by the programmer is known as Maskable interrupt.

20. What is Non-Maskable interrupts?
An interrupt which can be never be turned off (i.e., disabled) is known as Non-Maskable interrupt.

21. Which interrupts are generally used for critical events?
Non-Maskable interrupts are used in critical events. Such as Power failure, Emergency shut off etc.

22. What are the various segments registers in 8086?
Code, Data, Stack, Extra Segment registers in 8086.

23. What is the position of the Stack Pointer after the PUSH instruction?
The address line is 02 less than the earlier value.

24. What is the position of the Stack Pointer after the POP instruction?
The address line is 02 greater than the earlier value.

25. Logic calculations are done in which type of registers?
Accumulator is the register in which Arithmetic and Logic calculations are done.

5.2 CONTROL & COMMAND WORDS

5.2.1 PROGRAMMABLE SERIAL COMMUNICATION (8251)

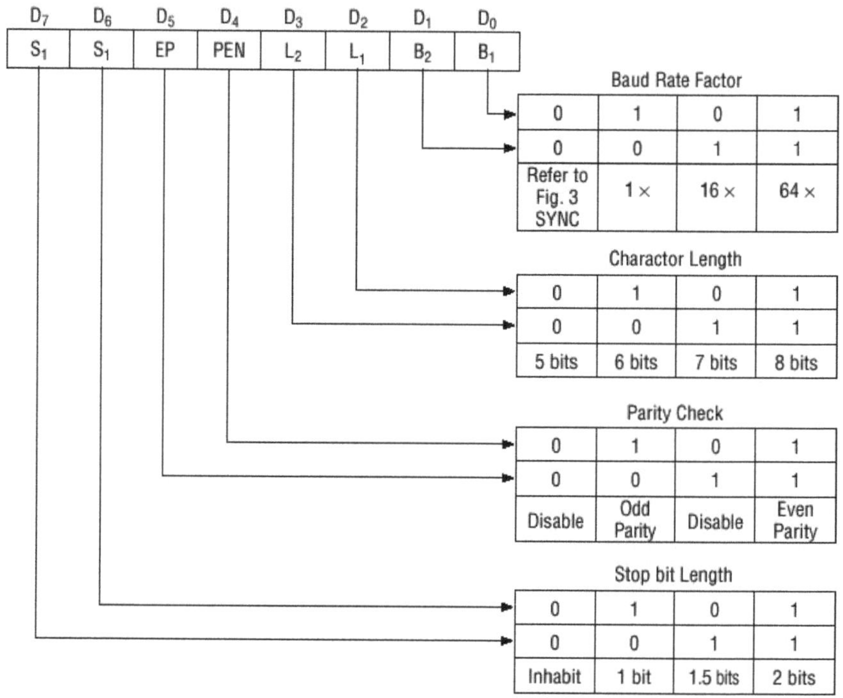

Bit Configuration of Mode Instruction (Asynchronous)

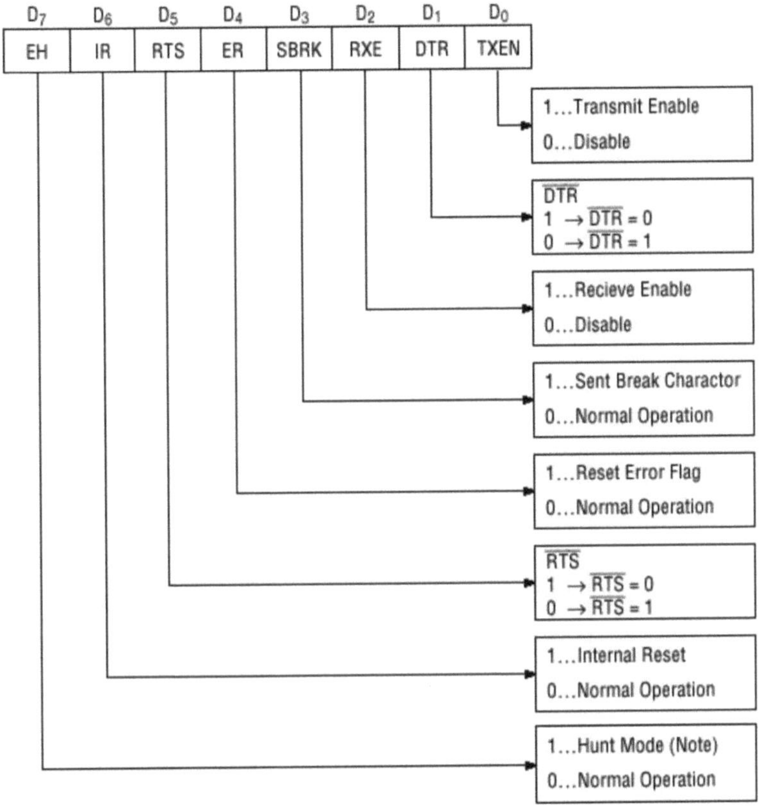

PROGRAMMABLE SERIAL COMMUNICATION (8251)

Bit Configuration of Command

5.2.2 PROGRAMMABLE PERIPHERAL PARALLEL COMMUNICATION INTERFACE (8255)

I/O MODE CONTROL WORD

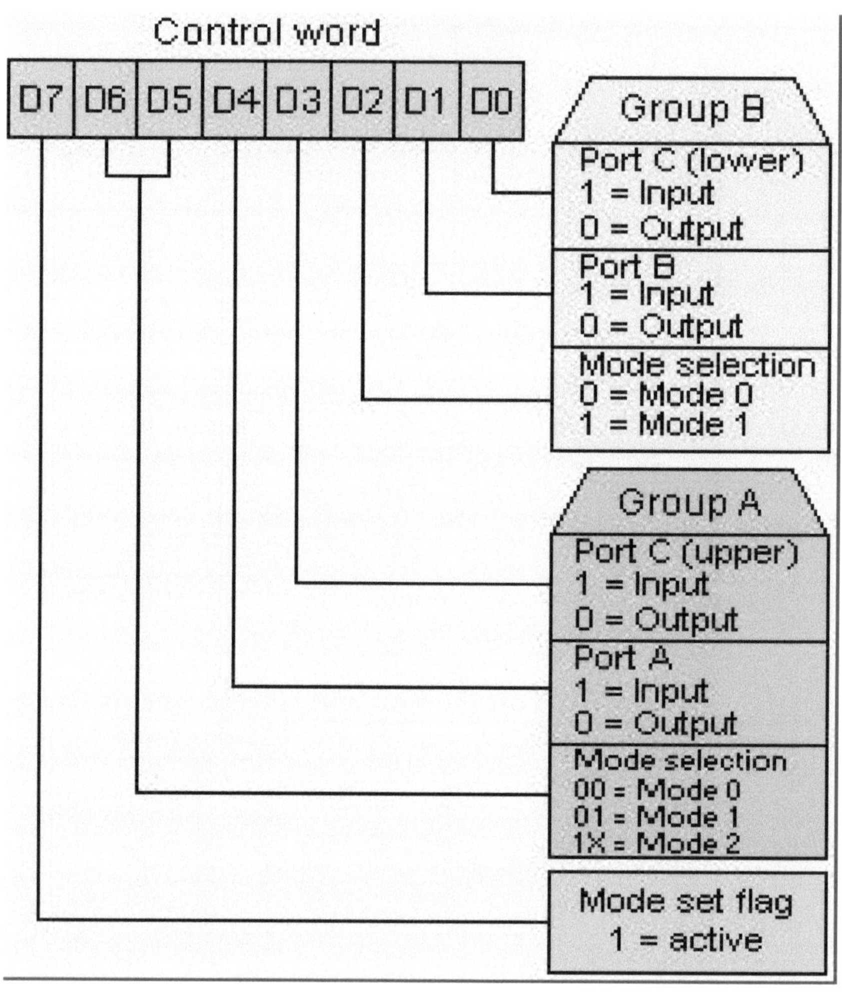

5.2.3 PROGRAMMABLE INTERVAL TIMER (8253)

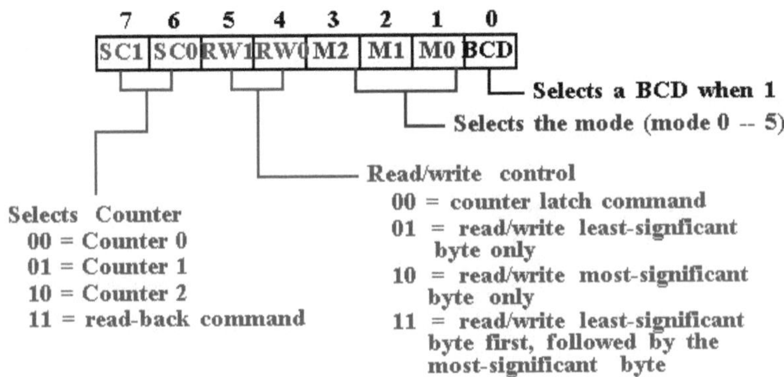

5.2.4 PROGRAMMABLE KEYBOARD & DISPLAY INTERFACE (8279)

1. KEYBOARD/DISPLAY MODE SET

0	0	0	D	D	K	K	K

DD – DISPLAY MODE
KKK - KEYBOARD MODE

kkk	Function
000	Encoded keyboard with 2-key lockout
001	Decoded keyboard with 2-key lockout
010	Encoded keyboard with N-key rollover
011	Decoded keyboard with N-key rollover
100	Encoded sensor matrix
101	Decoded sensor matrix
110	Strobed keyboard, encoded display scan
111	Strobed keyboard, decoded display scan

DD	Function
00	8-digit display with left entry
01	16-digit display with left entry
10	8-digit display with right entry
11	16-digit display with right entry

5.2.5 PROCEDURE FOR 8086 PROGRAMS USING MASM

1. Go to the start option, select RUN type CMD give enter
2. Type D: enter, type CD space your folder name give enter
3. Type EDIT give enter(using EDIT file type the 8086 programs)
4. After typing the program ALT F, Save as, Type file name.ASM,give enter, press ALT FX to exit from EDIT screen
5. Type MASM file name and press the enter key 4 times
6. Type LINK file name and press the enter key 4 times
7. Type DEBUG give enter, type n space bar file name.exe give enter,
8. Type L give enter, Type E 2000 give enter
9. Type the input data's by pressing the space bar key give enter
10. Type g=1000 give enter
11. Type E 2004 give enter
12. Type Q give enter

References

1. Yu-Cheng Liu, Glenn A.Gibson, ―Microcomputer Systems: The 8086 / 8088 Family - Architecture, Programming and Design‖, Second Edition, Prentice Hall of India, 2007.
2. Mohamed Ali Mazidi, Janice Gillispie Mazidi, Rolin McKinlay, ―The 8051 Microcontroller and Embedded Systems: Using Assembly and C‖, Second Edition, Pearson education, 2011.
3. Doughlas V.Hall, ―Microprocessors and Interfacing, Programming and Hardware‖,TMH,2012
4. A.K.Ray,K.M.Bhurchandi, "Advanced Microprocessors and Peripherals" 3 rd edition, Tata McGrawHill, 2012

yes
I want morebooks!

Buy your books fast and straightforward online - at one of world's fastest growing online book stores! Environmentally sound due to Print-on-Demand technologies.

Buy your books online at
www.morebooks.shop

Kaufen Sie Ihre Bücher schnell und unkompliziert online – auf einer der am schnellsten wachsenden Buchhandelsplattformen weltweit! Dank Print-On-Demand umwelt- und ressourcenschonend produziert.

Bücher schneller online kaufen
www.morebooks.shop

KS OmniScriptum Publishing
Brivibas gatve 197
LV-1039 Riga, Latvia
Telefax:+371 686 204 55

info@omniscriptum.com
www.omniscriptum.com

Printed by Books on Demand GmbH, Norderstedt / Germany